CLASS 47 AND 57 LOCOMOTIVES

CLASS 47 AND 57 LOCOMOTIVES

Ross Taylor

AMBERLEY

First published 2016

Amberley Publishing
The Hill, Stroud
Gloucestershire, GL5 4EP

www.amberley-books.com

ISBN 978 1 4456 5863 6 (print)
ISBN 978 1 4456 5864 3 (ebook)

British Library Cataloguing in Publication Data.
A catalogue record for this book is available from the British Library.

Typesetting by Amberley Publishing.
Printed in the UK.

Introduction

This book is written to provide a short history of the Brush/Sulzer Type 4 (Class 47) locomotive and the subsequent rebuilds of a small number of them to Class 57, illustrated with photographs from recent years.

Class 47

The Class 47 story goes back to the late 1950s, when British Railways – its Eastern Region in particular – was looking for a large build of Type 4 locomotives of a lighter weight than the existing lumbering English Electric and BR Derby 1Co-Co1 designs (classes 40 / 44 / 45 / 46). Brush of Loughborough was approached to design such a locomotive with an initial order for twenty. Following successful use in classes 44–46, power was to be provided by the Sulzer twelve-cylinder 12LDA28 diesel, in this case of the latest 'C' variant uprated to 2,750 bhp at 800 rpm. Subsequently, however, the use of this engine was to become the biggest weakness of the design, with stress cracks appearing in some crankcases after only a short period. As a result, all were de-rated to 2,580 bhp at 750 rpm in the mid-1960s, and a modification programme commenced. Generators, traction motors, etc. were all made by Brush.

The first locomotive D1500 (later No. 47401) was delivered in September 1962, by which time more had already been ordered, with subsequent orders taking the total up to 512. It concluded with D1961 (later No. 47515 *Night Mail*) in May 1968. As would be expected with such a large class, modifications were made during the course of delivery, one of the main ones being the wiring of the traction motors, which was changed from three pairs of motors connected in series (series – parallel) to all six in parallel on later locos. This was largely to reduce the risk of main generator flashover as it operates at half the voltage, and also to reduce wheel-slip problems as, in series parallel, if one axle slips the drop in load on that motor results in a voltage increase to the other paired motor, thus making it more likely to slip also.

As built, there were three main variants of the 'Brush/Sulzer' as it tended to be known, viz.:

- The initial twenty D1500–1519 (Nos 47401–420), fitted with both steam heating and DC Electric Train Heating (ETH) generator.
- Standard locos D1521–1781, 1837–1874, 1901–1999, 1100–1111 (Nos 47001–299 – but, by renumbering, many had already had ETH fitted and became

Class 47/4, D1960/1 being fitted from new and becoming 47514/5), fitted with steam train heating equipment.
- Freight locos D1782–1836, 1875–1900 (Nos 47301–381) built without train heating equipment; it was on these that the lack of wheel sanders across the whole class was most severely felt in times of poor adhesion.

Following delivery, the class was allocated to all BR regions apart from the Southern, and proved itself as a true mixed traffic locomotive, able to haul heavy freight trains but also operate express passenger services efficiently. One locomotive, D1733, was notable in being the very first to carry BR blue livery, although in an experimental form, as the loco used for the XP64 set of prototype coaching stock. On it the immortal 'arrows of indecision' logo made its debut, and happily this loco is still in existence as No. 47853, although at the time of writing its future is uncertain following the withdrawal of the entire Class 47 fleet operated by Direct Rail Services.

Mention should be made of the five locomotives (D1702–6) that were fitted at build on an experimental basis, with the new Sulzer 12LVA24 Vee engine of 2650 hp. This engine was 5 tons lighter than the standard unit, and also took up less space in the engine room, improving access. Under TOPS these locomotives became Class 48, but the fact that the engine proved to be of mediocre reliability, coupled with it being non-standard in the fleet, led to the decision to convert the locomotives to standard 12LDA28 engined Class 47s in 1969/70. One of these former Class 48s is preserved: D1705/No. 47117, which is normally to be found on the Great Central Railway at Loughborough.

The first major conversion to be undertaken commenced in 1969, following BR's decision to specify electric heating for future coaching stock, a decision that should have been made several years earlier, given that steam heating generators installed in diesel locomotives were an almost universal source of constant trouble, Class 47 being no exception. Over the years until 1986, standard 47/0s were steadily converted to supply ETH by replacing the auxiliary generator with a dual-wound alternator and adding the necessary control apparatus, being reclassified as 47/4 and ending with No. 47665, the former D1909.

In 1979, twelve locomotives were fitted with push-pull control equipment to operate an upgraded Edinburgh–Glasgow service with new Mk3 stock, replacing Class 27s on each end of a rake of early Mk2s, on which the constant high speeds were taking their toll. These conversions became the first 47/7s and, in 1985, four more were converted. The last was done in September 1988 as a replacement for the fire-damaged No. 47713. Following replacement by Class 158s in 1990, many of these saw work for Network South-East replacing Class 50s, which were being run down.

From 1986 a number of 47/4 locomotives were fitted with long-range fuel capacity by adding an extra tank where the boiler water tank had been, later being renumbered in the 47/8 series.

After sectorisation of BR in the late 1980s, locomotives were allocated to specific sectors and, in some cases, particular traffic. Those on freight work (47/0 and 47/3) were restricted to 75 mph to reduce stress on running gear, but those allocated to InterCity probably were worked harder than ever before. They were on constant high speed duties, with the resulting detrimental effect on reliability.

Privatisation saw most of remaining fleet owned by either: Porterbrook and leased to operator; by American-owned English Welsh & Scottish Railway, who took an instant dislike to them and replaced them as soon as possible with new General Motors Class 66 and 67; or by Freightliner, who made reliability modifications to their locos before developing the project to re-engineer some of them with replacement General Motors engines and upgraded electrical equipment, producing Class 57. These, and the later orders for Class 66s, saw the writing on the wall for their 47s.

Those owned by Porterbrook were operated by:

- Virgin, who ordered a fleet of Voyager DMUs to replace them by August 2002.
- First Great Western, who replaced theirs with Class 57s in 2004.
- 'Spot hire' companies, although some of these have come and gone. The first of these was music producer Pete Waterman's Waterman Railways, but he soon became disillusioned with the business and sold most of his stock. Several of the 47s passed to newly formed Fragonset Railways, who built up a substantial fleet of the type, although many were in need of repairs and never saw service. Fragonset merged with Merlin Rail in 2005, but the following year went into liquidation and the rolling stock was disposed of, with many of the 47s going for scrap.
- Cotswold Rail, who provided locomotives to Anglia and successors among other work. They too became bankrupt in 2011, but not before acquiring the short-lived company Advenza Freight, who had two Class 47s painted in a livery reminiscent of BR blue.
- Riviera Trains, who have operated a number of the class, recently supplying freight company GBRF with several locos for lighter duties to release that firm's Class 66s for heavy work. A number of Riviera's 47s (including No. 47853, the former D1733) were sold on to Direct Rail Services, who has recently withdrawn the last of the fleet following acquisition of more 37s and new 68s. One has to hope that at least some will find further work. West Coast Railway also has a large fleet of the type for use on charter work while Colas Rail own three used on freight.

To sum up the current position, eighty-two Class 47s survive, plus the thirty-three converted to 57s. Several are still in main-line operation, but mostly on a sporadic basis – a far cry from the days when they were to be seen virtually anywhere on the network. A number have been preserved and run regularly on heritage railways.

Class 57

Following the privatisation of Britain's railways in the mid-1990s, the freight operator Freightliner found itself in charge of a fleet of 47s of increasing unreliability. In addition to commencing power unit overhauls and other improvement work, it was realised that newer traction was required. Rather than going down the road followed by EWS, who ordered 250 new Class 66s from General Motors in America, Freightliner, in conjunction with Porterbrook Leasing and Brush, pursued a re-engineering programme for some of its 47s. This plan had originally been devised by Brush for Railfreight distribution (including new Class 60 style cabs, subsequently abandoned on cost grounds). This programme produced a locomotive that was, in many ways, new but at a reduced cost and also in a shorter timescale than building new. The first of these, No. 57001 (the former No. 47356), entered traffic on 21 July 1998 and soon proved a success. The conversion work included:

- Replacement of the Sulzer 12LDA28C diesel engine, with a fully remanufactured GM 645-12E3 2 stroke – this engine was a twelve-cylinder version of the highly successful sixteen-cylinder one used in the Class 59, and was of proven reliability.
- Replacement of the original main generator with a reconditioned and modified ex-Class 56 alternator, feeding a Class 60-type rectifier.
- New wheel slip protection was fitted and sanding gear finally fitted.
- Internal refurbishment of cabs and an overhaul for all other existing equipment.

Over the following eighteen months a further eleven locomotives were converted, although at one point there were plans for a fleet of thirty, subsequently reduced in favour of new Class 66s.

Given the success of these locomotives, Porterbrook had a further one converted in 2001, but with the addition of ETH, with a view to an order for passenger work. This became No. 57601 (formerly No. 47825) and featured a cab that had been tidied up considerably, all traces of the former headcode box being removed and new light units being fitted. In fact, it looked distinctly similar to the locomotives supplied to Cuba in 1963, which were essentially Class 47s (though with the 12LVA24 engine). It was painted in an eye-catching purple/silver livery, with full height numbers, and commenced work for First Great Western on the London–Penzance route. It generally performed well and, following talks, Porterbrook leased four locomotives to FGW – Nos 57602–5, which entered traffic in 2003/04. These locomotives had a different version of the GM twelve-cylinder engine; in this case, the 645-12F3B of 2750 hp, the extra output offsetting the power drawn by the ETH supply. At the time of writing, the locomotives continue to perform the work for which they were ordered, this mostly being the sleeper service between London and Penzance. Once the production 57/6s were imminent, No. 57601 became non-standard and was sold to West Coast Railway company, who continue to use it on charter work.

Also as a result of the success of No. 57601, Virgin ordered sixteen ETH-fitted locomotives to become 57301–316 for use in hauling Pendolino units either when they failed, e.g. if the overhead power was off, or on non-electrified routes, mostly as diversions when the main line was closed for engineering, but also to allow through services to destinations off the main electrified route. The 57/3 had a similar specification to the production 57/6 and were built between 2002 and 2004. The last three locomotives had a Dellner coupling fitted from new to allow simple coupling to the Pendolino sets, while the remainder of the fleet was retrofitted with the couplers. This coupling is retractable, and required substantial modification to the front of the locomotive, although it ruined the tidy appearance created following rebuild.

Today, all the Class 57s remain in service but, with the exception of the Great Western ones, all have changed operators and duties. The first to change hands were the 57/0s as, following the purchase of a considerable number of new Class 66s, Freightliner no longer required the 57 fleet. During 2007, the entire fleet came off lease in two batches, the first of which, Nos 57006–012, soon found a new home with DRS, who also took three of the second batch stood down by Freightliner. Two (Nos 57005/6) went to Advenza Freight in 2008, however. Following the liquidation of this company, they passed to WCRC, together with pioneer No. 57001, which had previously been stored and used as a 'Christmas tree' at Brush for other locomotives.

By 2008, Virgin was finding it no longer had sufficient work for its fleet of 57/3s, and four were subleased to Arriva Trains Wales to operate their new Cardiff–Holyhead loco-hauled service. These were Nos 57313-6, 314/5 being painted in full ATW livery, while the other two became plain blue as back-up locomotives for the ATW service but also performed Virgin work. The 57s operated this train until 23 March 2012, when Class 67s took over, and the 57s subsequently sold to WCRC. Further spare capacity in the Virgin fleet in 2011 saw six locomotives leased by Network Rail mainly for work south of London; these were painted in NR yellow livery, but were then sold by Porterbrook (who still owned them at this point) to DRS in 2014. There they joined six of their sisters, which had already been leased to the operator, though these latter were hired to Virgin for standby Thunderbird duties. The DRS locos perform a variety of work, including the prestigious Northern Belle for which Nos 57305 & 312 are painted.

This to date, in 2016, is the story of the Class 47 and 57. The 57s undoubtedly will be around on the main line for some years to come, but the 47s are very much in the twilight of their careers now, and I would expect it not to be very much longer before the only main-line work will be by preserved examples on charters. However, bearing in mind that, in the 1980s, British Rail was regarding them as having a thirty-year life, which would have seen them all withdrawn by the turn of the century, the fact that some have clocked up fifty years' service is testimony to their design.

SRPS No. 47643 is seen arriving into Ramsbottom while in charge of the 11.15 2J60 Rawtenstall–Heywood service. Seen here on Saturday 4 July 2015. The locomotive is based at the Bo'ness & Kinneil Railway, but was one of the many visiting locomotives for the 2015 summer diesel gala event.

Fresh from the paint shop is ex-Royal No. 47798 *Prince William*, seen here on hire to West Coast Railways to shunt the 5Z69 13.36 empty stock into Holgate sidings. Seen here awaiting time on platform 10 at York station on 12 March 2015.

Colas Rail have three Class 47s in operation: Nos 47727, -739 and -749. Here we see No. 47739 *Robin of Templecombe* supporting a 'Ram' headboard as it storms northwards at Winwick Junction, while working 6Z42 Chirk Kronospan–Ribblehead Virtual Quarry log empties. Seen here on 14 August 2010.

DRS No. 47501 *Craftsman* departs from Workington North with the 2Z33 Floodex service from Maryport. No. 57012 was on the rear end of the train, seen here on 12 April 2010.

Crewe Heritage Centre ran three return trips over the Weardale Railway over the weekend of 5 and 6 February. Here we see ScotRail-liveried No. 47712 heading in the Bishop Auckland direction at Witton-le-Wear, with a service from Stanhope. No. 47192 was on the other end of the train. Seen here on Sunday 6 February 2016.

No. 47812 leads classmate No. 47843 *Vulcan* while in charge of the 4D93 Doncaster Down Decoy–Drax power station gypsum containers. Seen here passing Southmoor Lane on the outskirts of Knottingley on 3 November 2014. In the distance, Ferrybridge power station can be seen.

DRS-owned No. 47832 *Galloway Princess* sports the Northern Belle livery as it powers down the East Lancashire Line at Hapton while in charge of the 1Z35 12.28 Blackpool North–Wakefield Kirkgate VSOE Northern Belle £195 Luncheon. Seen here on 25 September 2011.

Freightliner-liveried No. 47292 sits at Crewe Basford Hall, after just being put into the DHLT storing pool. The locomotive has since been bought and now lives at the Great Central Railway in Ruddington. It was photographed here on 15 August 2004.

Following on from the previous photograph, here we see No. 47292, looking rather splendid this time as it sits in the sun at Ruddington. The locomotive was withdrawn on 28 February 2007 after being purchased but was reinstated on 31 July 2009. Seen here on 6 September 2015.

The EWS yard at Warrington Arpley was once busy with variety, as seen here in this picture: No. 47786 *Roy Castle OBE* sits among other classes such as the 08, 37, 60 and 66. The Class 47 had been in a little accident, hence having no windscreen in one half of the front cab. Photographed here on 8 April 2004. The locomotive is still in use for West Coast railways at Carnforth.

The only West Coast Railways Class 47 to survive being painted is No. 47270, seen living up to its name as it *Swift*-ly rounds the curve at Hest Bank with the 5Z18 Carnforth Steamtown–Crewe C.S empty stock working. Seen here on 30 August 2012.

Direct Rail Services No. 47805 leads the 1Z80 Southampton Docks–Glasgow Central Cruise Saver past the newly built Buckshaw Parkway station. Photographed here on Sunday 17 June 2012. Classmate No. 47802 was on the rear of the train.

Northern Belle-liveried No. 47832 leads the 1Z94 Edinburgh Waverley–London Euston Northern Belle through Red Bank near Newton-le-Willows. DRS-liveried No. 47813 was on the rear and is seen here on 19 August 2013. The Class 47 is now owned by West Coast Railways at Carnforth, due to DRS withdrawing all their Class 47s.

Colas Rail was in charge of the 6Z92 Crewe Electric Depot–Dollands Moor locomotive move and was captured at Bathpool. Crewe Works-built (February 1965) No. 47749 clags upgrade from Kidsgrove towards Stoke-on-Trent, with long-term Nos 92033, -006, -018, which were bound for GBRF. The two steel coil carriers were acting as brake force while Crewe Works-built (October 1964) No. 47727 idles, bringing up the rear-fitted portion. Seen here on 13 July 2011.

West Coast Railways No. 47826 *Springburn* sits at Hellifield, awaiting the arrival of LMS Coronation Class 4-6-2 No. 46233 *Duchess of Sutherland*, which was hauling the 1Z63 'The Cumbrian Mountain Dalesman'. The Class 47 was waiting to take over the rail tour, which was heading back to Kettering. Seen here on 5 August 2006.

Crewe Heritage Centre ran three return trips over the Weardale Railway over the weekend of 5 and 6 February. During the second trip the locomotives were run round to have a change of climbing up the gradients in the Weardale valley. Here we see No. 47712 sitting next to classmate No. 47192 for a photograph shoot at Stanhope.

No. 47832 sits in Crewe works as a star guest to the 2005 open day event. The locomotive is seen supporting FM Rail colours, who were a railway charter company based in Derby. The company was formed when Fragonset Railways and charter operator Merlin Rail merged together. The company went into administration in December 2006. Photographed here on 11 September 2005.

The Midland Railway Centre at Butterley is home to D1500 (No. 47401) *North Eastern*. The locomotive is owned by the 47401 Project, who purchased the locomotive back in early 1993. Here we see it curving away from Swanwick Junction while working the 10.00 Butterley–Riddings service. Seen here on 9 April 2006.

Direct Rail Services No. 47805 leads the 1Z68 Carlisle–Manchester Victoria day four of 'The Africa Express' past Hest Bank near Morecambe in Lancashire. Seen here on 5 September 2012.

On hire to GBRF, Class 47s Nos 47843 *Vulcan* (leading) and No. 47812/D1916 fast approach Joan Croft Junction, while in charge of the 4D31 16.08 Drax power station–Doncaster Roberts Road loaded gypsum containers. Seen here on 26 September 2014.

EWS No. 47760 *Ribblehead Viaduct* sits at Sellafield, waiting to work the return leg of the 15.40 1Z71 Ravenglass–Norwich Nenta rail tour 'The Cumbria & Eskdale Explorer'. Seen here on 19 April 2003.

After having engine repairs at Carnforth, national treasure ex-Res/EWS Royal train loco No. 47798 *Prince William* worked several charters for West Coast Railways. It's seen here returning the empty stock via the 'Little' North-Western Railway back to Carnforth from a previous day's charter. The working was the 5Z47 Barnetby–Carnforth and was seen here on 12 April 2009.

Looking rather derelict and forgotten is Freightliner No. 47370 as it is seen here, neglected, at Ipswich. The locomotive had been in long-term storage due to the newer Class 66 having taken over its booked workings. It was seen here on 29 March 2007 and was disposed at T.J. Thomson in Stockton on 5 August 2009.

After being stored at Norwich Crown Point in January 2008, No. 47818 was then moved to Barrow Hill in preparation of its new life with Direct Rail Services. The locomotive is seen holding its ex-Greater Anglia colours as it is used for running shuttles. Seen here on 8 August 2009.

Brush type 4 No. 47643 powers past Burrs country park while in charge of the 2J66 13.33 Rawtenstall–Heywood service. Seen here on Friday 3 July 2015. The Class 47 has come from the Bo'ness and Kinneil Railway, and was one of the visiting locomotives along with Nos 26038, 20020, 37025, and 31601.

Colas Rail No. 47739 *Robin of Templecombe* passes Taylors Bridge in Clitheroe while working the 6Z41 14.10 Ribblehead Virtual Quarry–Chirk Kronospan loaded log train. Seen here on 13 October 2010.

Vintage Trains No. 47773 (D1755) coasts past Langho on the Ribble Valley Line with the 1Z45 Tyseley Steam Trust–Carlisle 'Cumbrian Guardsman' rail tour. Seen here on Saturday 2 March 2013.

Freightliner green-liveried No. 47150 supports the East Lancashire Railway as a guest locomotive for the annual diesel gala. The Freightliner engine was captured here departing Irwell Vale in style, while in charge of the 09.00 Bury–Rawtenstall service. Seen here on 14 September 2003. The locomotive has now been scrapped, which took place at T.J. Thomson in Stockton on 31 October 2010.

Double duff delight! Nos 47815 *Great Western* and 47843 *Vulcan* speed through Knottingley with the 4D93 1028 Doncaster Down Decoy GBRF to Drax Aes (GBRF) empty gypsum containers in tow. Seen here on 17 October 2014. Ferrybridge power station is visible in the background.

DRS No. 47501 *Craftsman* anchors on for Crewe with 5Z49 Preston PCD–Crewe Holding Sidings ECS move. The train had worked 'The Devonian' charter for Compass Railtours the previous day, from Preston to Paignton and return. The train is formed of thirteen Mk2s with No. 47712 *Pride of Carlisle* bringing up the rear. Seen here passing Coppenhall on 29 August 2010.

With a toot from DRS driver Ian Tunstall, Nos 47805 *John Scott* and 47853 *Rail Express* head the 1Z72 'Spring Scenic Settler' rail tour from Carlisle to Swindon through the beautiful Ribble Valley near Clitheroe. The train is pictured on the return journey back to Swindon, seen here on 17 April 2014.

The West Coast Railways maroon makes the picture look even more dull as No. 47245 sits with No. 47237 at Hellifield goods loop, awaiting the arrival of 8F 48151. Seen here on 5 June 2014.

Riviera Trains No. 47843 *Vulcan* storms through Clitheroe with the 1Z35 'Edinburgh Festival Fellsman', which was running from Stoke-on-Trent to Edinburgh Waverley. Seen here on 28 August 2009.

EWS No. 47812, still carrying Virgin livery, climbs up the gradient away from Manchester Victoria station as it passes Miles Platting Junction, while working empty rail carriers. The working was the 6J65 06.24 SX Crewe Basford Hall Sorting Sidings North–Castleton Rail welding depot. It was seen here on 18 May 2005.

Cotswold Rail-liveried No. 47200 *The Fosse Way* sits at Preston PCD sidings, awaiting its next duties. The locomotive was seen here on 20 July 2005. It was stored several times after being seen here, but ended its days some three years later at T.J. Thomson's at Stockton in May 2008.

Seen inside the shed at EWS Wigan Springs Branch depot (otherwise known as Wigan CRDC), No. 47312 meets the end of its life as it is slowly being broken up. Photographed on 31 March 2003.

On 9 August 2012 Colas-owned No. 47739 *Robin of Templecombe* passes Barrow-on-Trent with the 6E07 Washwood Heath to Boston Docks steel train.

Preserved No. 47105 operates a demonstration freight to Cheltenham Racecourse at Hailes on the Gloucestershire Warwickshire Railway. This loco was new in December 1963 as D1693. Seen here on 2 April 2005.

No. 47580 *County of Essex*, in the very popular BR Large Logo livery, attends Eastleigh Works open day on 23 May 2009.

Withdrawn Freightliner no-heat No. 47302 awaits its fate at Crewe Basford Hall on 1 June 2003. It was finally reduced to scrap metal on 25 August 2007 at T.J. Thomson, Stockton.

DRS-liveried No. 47802 *Pride of Cumbria* takes a rest at Preston after powering down the WCML from Carnforth. The working was 1Z40 Leeds–Leeds Northern Belle Circular via Hellifield and Copy Pit on 29 September 2011. Delivered in 1966 as D1950, she was renumbered 47552 under TOPS before becoming a long range fuel capacity locomotive and being renumbered again as 47802.

One can almost feel the cold as two Brush Type 4 Nos 47839 and -853 stand on Crewe Diesel depot on 22 December 2010.

Revised Network SouthEast-liveried No. 47596 *Aldeburgh Festival* makes a striking sight on 2 July 2013 on the tail of a Rawtenstall service at Heap Bridge on the East Lancashire Railway. It was visiting for the diesel gala and is normally based at the Mid-Norfolk Railway.

Stanhope station on the Weardale Railway is the location for this picture of No. 47192. It also carries its old number 1842 on one side. A number of 47s ran in service in green livery, carrying TOPS numbers, in the 1970s prior to being repainted in corporate blue. The service was the 12.30 to Primrose Hill on Sunday 6 February 2016.

Over the weeks in the run up to Christmas 2014, Northern Rail operated extra loco-hauled trains to convey shoppers into Manchester. They were operated by DRS 47s and Mk2 stock – and on 22 November No. 47790 in Northern Belle livery works 2Z13 Manchester Victoria–Preston at Buckshaw Parkway. Sister No. 47818 was on the rear.

DRS No. 47810 *Peter Bath MBE* is seen passing through Leyland, after being rescued by classmate No. 47790 *Galloway Princess* after failing with the 1Z17 Bradford Interchange–Blackburn extra on 7 July 2014. Several additional trains were operated over this weekend to cope with demand during the passage of the Tour de France bicycle race.

Before entering preservation No. 47306 stands at St Blazey depot after its service days were over on 29 November 2008. It had been stored in 2002, still carrying triple-grey Railfreight Distribution livery, which it carries today in preservation at the Bodmin & Wenford Railway.

Owned by the Stratford 47 group, main-line-registered No. 47580 passes Settle Junction as it stands in for steam traction on West Coast Railway Company's 'The Waverley' rail tour, which was the return working of the 1Z71 Carlisle–York, seen on 20 July 2014. Diesel haulage was required due to a ban on steam resulting from the prevailing dry weather and risk of lineside fires being caused from steam locos.

Fragonset No. 47703 *Hermes* passes Horrocksford Junction, Clitheroe from Layton near Blackpool to Edinburgh Waverley with the 1Z54 'The Edinburgh Fellsman' on 23 September 2006. No. 47703 would be no stranger to Edinburgh, having been one of the locomotives converted for push-pull working of the Edinburgh–Glasgow service in 1979, when it was named *St Mungo*.

Burton Salmon is the location of preserved No. 31271 and GBRF London Underground-liveried No. 66721, being towed by No. 47843 *Vulcan* as the 0Z31 Wensleydale Railway–Midland Railway, Butterley on 21 July 2014.

No. 47270 *Swift* heads 5Z27 15.00 Carnforth–York ECS, which was to form the following days' Scarborough Spa Express into a rain-soaked Skipton on 18 July 2011. LMS 5MT was the traction for the tour itself.

Twin fuel tank No. 47810 crosses Docker Viaduct with 1Z33 06.10 Warrington Bank Quay–Helensburgh Upper Northern Belle. No. 47790, in the dedicated livery for these trains, is on the rear, seen on 3 June 2011.

On 26 November 2013, No. 47811 was being used for duties far removed from its InterCity days, as it was operating for Freightliner as a yard shunter at Crewe Basford Hall.

Carrying former operator Virgin's livery, Freightliner No. 47812 passes Hellifield South Junction box with a Derby–Carlisle rail-grinding train on 21 July 2003. New as D1916, this loco became No. 47239 under TOPS before being one of the last to undergo ETH conversion in 1986 to become No. 47657. It received long-range fuel tanks at the same time as ETH conversion, in common with the other late conversions, Nos 47650–665, and became No. 47812 in 1989. It carried the name *Pride of Eastleigh* for eighteen months from February 2002, remaining in main-line service today with Riviera Trains.

Crewe Diesel depot is host to XP64-liveried D1733/47853 and No. 47848 on 14 August 2010.

Following overhead wire damage, due to high winds on the Midland main line, the OHLE train was hauled from Preston Dock Street to Bedford by No. 47826 *Springburn*. Here, it waits at Preston station, platform 7, on 27 March 2003.

Colas No. 47727 *Rebecca* is hauled by Class 60 No. 60087 through Doncaster station on 3 March 2015 as the 0Z56 Doncaster CHS–Tyne SS.

On 16 May 2014, No. 47245 hauls failed Colas No. 66847 past Slindon on its way from Cardiff Canton to Longport EMD.

Pioneer No. 47401 *North Eastern* stands at Swanwick on the Midland Railway, Butterley, on 18 March 2015. Delivered way back in September 1962 as D1500, she spent most of her life as an Eastern Region engine, working East Coast main-line expresses for several years. Preserved in 1993, it now leads a rather more gentle life on heritage railways.

Riviera-owned No. 47815 *Great Western* is seen on hire to GBRF, working 6Z47 York GR shed–Doncaster Roberts Road, passing Burn on the ECML. Pictured on 7 August 2014. The 'Great Western' name was formerly carried by No. 47500, which was one of the four Class 47s painted in GW green for the GW150 celebrations in 1985.

No. 47790 *Galloway Princess* top-and-tails with Class 57 No. 57302 on the 1Z74 Manchester Piccadilly–Crewe Northern Belle working, and is pictured passing through Warrington Bank Quay. Dated 21 September 2014.

The second of the original batch of 'Generators', No. 47402, is pictured at Ramsbottom on the East Lancashire Railway, with later loco No. 47575 on 9 July 2006. Sadly, the latter was scrapped in May 2010, having had a life of almost forty-six years after being delivered as D1770 in 1964.

No. 47501 *Craftsman* accelerates away from Flimby with the 2Z27 Maryport–Workington service on 12 April 2010. No. 57012 is on the rear. This service was one of the short-notice trains operated by DRS that were provided following the floods in Workington in late 2009, which severely damaged the road bridges in the town, effectively cutting the town in two. For several months, these trains were the only link between the two parts of the town.

Nos 47843 and 47812 depart from Drax power station a little earlier than the booked time of 12.28 while working the 4D19 Drax power station–Doncaster Down Decoy GBRF loaded gypsum train. Seen here on 7 August 2014.

Also on 7 August 2014, Nos 47843 and 47812 pass Burn with the 4D19 12.28 Drax power station–Doncaster Down Decoy loaded gypsum train. These locomotives are hired by GBRF from Riviera for these trains, in order to release Class 66s for heavier duties elsewhere.

No. 47818 sits at Preston, after arriving with the 3J11 Carlisle Kingmoor–Carlisle Kingmoor RHTT working. Normally this train is hauled by Class 66s but, due to a shortage, the 47 was substituted. Seen here on 27 November 2013.

No. 47643 sits at Bury Bolton Street, awaiting its timetabled departure of 17.00 to Heywood. The locomotive was one of many guest locomotives for the 2015 diesel gala and is seen here on 4 July 2015.

No. 47501 heads and No. 47805 *John Scott 12.5.4522.5.12* tails as the 1Z30 Crewe–Hellifield Northern Belle passes Taylors House on 18 December 2013.

DRS No. 47853 *Rail Express* and No. 47501 stand on shed at Crewe Gresty Bridge depot on 9 December 2013. By this time -853 was in standard DRS livery, with no indication that it was the former D1733 XP64 loco.

On 26 July 2013, D1916 /47812 in the original two-tone green livery, but with yellow ends, shunts Mk1 coaching stock that was to be used for a GBRF rail tour the following day. Bristol Temple Meads is the location of this picture where, judging by the flora growing in the track bed, a visit from the weed-killing train is overdue!

Its working days now over, No. 47287 stands forlornly at Crewe Diesel depot on 19 August 2004. Delivered in February 1966 as D1989, it survived for almost thirty years, being finally cut up at CF Booth, Rotherham in January 2005.

No. 47501 powers past Taylors Bridge in Clitheroe with 1Z62 06.46 Crewe–Ravenglass Northern Belle. No. 47818 was on the rear on 1 May 2013.

No. 47802 leads -298 as they pass Farington Curve Junction on the outskirts of Preston while working the 4Z26 Stoke Marcroft–Carlisle Kingmoor wagons from repair train.

One of the first batch of push-pull 47/7 conversions, No. 47712 passes Kingfisher Park on the outskirts of Frosterley, with the 12.30 Stanhope–Primrose Hill, as it stars in a Class 47 running day at the Weardale Railway, held in conjunction with Crewe Heritage Centre. It wears the attractive ScotRail version of InterCity livery, as applied in the late 1980s, photographed on 13 February 2016.

Driver Ian Tunstall puts on a little show for the camera as he applies power on Nos 47805 and 47853, passing Taylors House in Clitheroe with the 1Z72 Carlisle–Swindon Pathfinder rail tour on 17 May 2014. Much of the stock are Mk2s, formerly used by the now-defunct FM Rail for their 'Blue Pullman' trains, and they carry the Nanking Blue livery applied to the Blue Pullman diesel units in the 1960s.

In a livery based on that carried by Nos 47163/4 for the Queen's Silver Jubilee in 1977, No. 47580 passes Burrs country park on the East Lancashire Railway on 5 July 2014. While most of the stock is BR Mk1, the vehicle immediately behind the locomotive is a Great Western auto-trailer, which would have been more commonly working with a 14xx GWR tank loco.

No. 47813 hauls classmates Nos 47712 and 47818 past Lancaster as a Carlisle Kingmoor–Crewe Gresty Bridge working, these being the two DRS depots. Photographed on 23 July 2015.

WCRC Nos 47832 and 57314 are stabled at Hellifield on 28 July 2015.

On 28 November 2008, Cotswold Rail Nos 47714 and 47701 await their next duties at Gloucester depot, in company with two earlier Brush products in the form of two ex-Fragonset Class 31s. No. 47714 is in Anglia Railways livery, for whom Cotswold carried out a considerable amount of work.

On 21 August 2014, the BR green livery on No. 47812 is starting to look faded as she passes through Doncaster with 4D93 09.41 Doncaster Down Decoy–Drax gypsum empties.

The now-scrapped No. 47839 passes Wilpshire with Compass Railtours 1Z48 'The Cumbrian Fellsman' Carlisle–Stafford tour on 18 April 2009. Delivered in 1964 as D1728, it became No. 47136 under TOPS renumbering. Following ETH conversion in 1984, it was renumbered again to 47621 before becoming 47839 in late 1989 when long-range tanks had been fitted. August 2013 saw its appointment with the cutters' torch at Eastleigh.

In 1980s Large Logo livery, No. 47847 leads standard BR blue No. 47843 *Vulcan* as the 4D31 Drax power station–Doncaster Down Decoy loaded gypsum train on 13 April 2015. No. 47843 acquired its name when just six months old as D1676 in October 1965, subsequently becoming No. 47090 under TOPS, then No. 47623 when converted to a Class 47/4 in November 1984, before having twin tanks fitted and becoming No. 47843 in the last month of the 1980s.

WCRC No. 47760 stands under Wainwright bridge, Blackburn, prior to working the ECS for the 'Tin Bath' rail tour to Preston on 13 February 2016.

No. 47192 sits in Walsingham depot on the Weardale Railway, awaiting its next duty on the Polar Express service on 29 November 2015.

Minus its offside windscreen, No. 47786 awaits repair at Warrington Arpley yard. This locomotive is now operated by WCRC, and carries their standard livery.

With a mixed rake of Mk2 coaching stock, No. 47501 passes through Carnforth station with the 1Z50 Carlisle–Hooton 'Settle–Carlisle Express' rail tour on 11 September 2013. Carnforth station was made famous by the 1945 film *Brief Encounter*, part of which was filmed there.

Following on from page 32, here is another view of DRS No. 47810 *Peter Bath MBE*, which is seen sporting DRS's latest livery as it passes through Leyland after failing with the 1Z17 Bradford Interchange–Blackburn extra on 7 July 2014. Several additional trains were operated over this weekend to cope with demand during the passage of the Tour de France bicycle race.

Resident No. 47449 (formerly D1566) runs round its train at Llangollen after arriving with a service from Carrog on 17 March 2007.

Deep in the Scottish Highlands, WCRC No. 47854 is near Spean Bridge with the 1H86 Mallaig–Crianlarich 'Royal Scotsman' luxury train. Some of the coaching stock is the Metro-Cammell Pullmans once used on the Eastern Region.

Seen on the rear of the 0K05 is Northern Belle-liveried No. 47790 *Galloway Princess*. Owned by DRS, the loco was being hauled from Carlisle yard to Crewe Basford Hall, ready for the next day's Northern Belle train. Seen here passing Clitheroe station with sister No. 47828 and Class 66s 66429/303 on 19 September 2014.

Freightliners' No. 47830 passes Farington Curve Junction, working 6Z52 MOD Longtown–Crewe Basford Hall consisting of Mk3 DVTs that, once overhauled, were bound for Norwich–London services.

York parcel sidings play host to DRS No. 47712, which awaits its next duty on 24 January 2009. One of the former ScotRail push-pull 47s, it is now preserved in ScotRail livery.

The famous D1733/47853 stands under the station lights at Preston at the head of 2Z29 from Manchester, another of the extra services provided by Northern Rail to cater for Christmas shoppers in 2014. Photographed on 13 December 2014.

No. 47829 is pictured at Leeds Midland Road depot. By this time it was operated by Freightliner, but its previous operator Virgin had painted it in police livery in March 2002 as part of a campaign to target railway trespass and vandalism. This is certainly the most garish livery ever carried by a Class 47, and a far cry from the two-tone green it carried when new as D1964 in September 1965. It was not the first time a locomotive had been painted in a police livery, however, as Class 37 No. 37093 had been done for a television advert in the 1980s.

Passing through the beautiful Ribble Valley are Direct Rail Services Nos 57003 and 57007 as they head the 6K05 Carlisle Yard–Crewe Basford Hall departmental train through the outskirts of Clitheroe. Seen here on 10 March 2014.

No. 57002 *Freightliner Phoenix* powers its train past Whitacre Junction while working the 4L90 0912 Lawley Street FLT–Felixstowe loaded container service. Seen here on 7 September 2006.

The North Yorkshire Moors Railway held the annual diesel gala over the weekend of 12 and 13 April 2003. As a star guest, Virgin Trains sent freshly painted Thunderbird No. 57306 *Jeff Tracey*, and it was photographed here at Pickering, waiting to work a service along the line.

Network Rail Thunderbird No. 57312 speeds over the River Soar on the Midland Mainline as it is seen running as 0M57 Eastleigh Works–Derby RTC light engine move. Seen here passing Normanton on Soar on 6 September 2012.

Direct Rail Services (DRS) owned No. 57305 *Northern Princess* heads the 1Z72 1658 Oxenholme Lake District–Cardiff Central Northern Belle as it passes through Hest Bank near Morecambe bay. DRS liveried No. 57301 was on the rear of the train. Seen here on 28 August 2015.

Now owned by West Coast Railways, No. 57006 was seen here sporting Advenza Freight livery. The locomotive is seen at Barrow Hill on 8 August 2009.

West Coast Railway Company 5Z47 Saltburn–Carnforth Steamtown empty stock move starts the climb out of Skipton towards Hellifield at Niffanys with 'body snatcher' No. 57601 leading No. 47804, which was tailing the formation. The Class 57 is the former No. 47825, once named *Thomas Telford*, which started its career on 25 August 1964 and was built at Brush Falcon Works Loughborough. In 2000, the Class 47 returned to Brush Traction for a Porterbrook Leasing trial, which involved totally stripping it down and rebuilding/rewiring its body, while its Sulzer power unit was removed and replaced with a reconditioned GM 2stroke V12 power unit. The generator was also replaced with a reconditioned Class 56 alternator and the ETS supply was uprated for suppling long and power-hungry sleeper coaching stock trains. The result of this prototype loco lead First Great Western to order four further examples; on delivery, No. 57601 became redundant and West Coast Railway snapped it up.

Running around twenty-six minutes late, Nos 57314 (47372) and 390025 depart Rhyl with the 1D83 London Euston–Holyhead passenger service. Seen here on Saturday 5 May 2012. The Pendolino works under its own power to Crewe before getting attached to the Class 57.

Virgin Trains Thunderbird No. 57304 *Gordon Tracy* drags Pendolino No. 390018 through Penmaenmawr on the North Wales coast, while working the 1D83 London Euston–Holyhead passenger service. Seen here on 29 January 2011.

After winning the contract of Freightliner, the 6K27 1443 Carlisle NY–Crewe Bas Hall SSM is now worked by rival Direct Rail Services. The train is seen here with No. 57009 leading, as it waits for the signal to change at Preston. It was seen here running some 140 minutes early on 7 July 2014.

West Coast Railway No. 57314 leads the 5Z67 Carnforth Steamtown–Scarborough empty stock through Giggleswick near Settle. Ex-Arriva Trains Wales blue No. 57313 was on the rear. Seen here on 22 November 2013.

Nos 57310, 68012 and 57312 work in top-and-tail mode while dragging a very poorly brand-new Chiltern Railways-liveried Class 68 back to Crewe Gresty Bridge from Wembley LMD. Seen here on 16 May 2015.

Double-headed Nos 57009 and -012 power away from Crewe Basford Hall Junction with the 6Z65 08.42 Crewe Coal Sidings–Avonmouth special. This is formed of the two ex-works DRS Escort Saloons 9428 and 9419, plus the two huge MOD flask carriers Nos 95770 and 95771. Seen here on 11 March 2014.

First Great Western own four class 57/6s and here is one of them, No. 57604 *Pendennis Castle*, seen on Long Rock depot in Penzance being cleaned after arriving with the 1C99 from London Paddington. It was seen here on 4 October 2011.

Ex-Arriva Trains Wales blue-liveried No. 57316 powers past Taylors House in Clitheroe while working 'The Settle & Carlisle Statesman' 1Z69 15.35 Carlisle–Hereford charter service. It was seen here on 31 May 2014. No. 57313 was on the rear.

Virgin Trains Thunderbird No. 57313 *Tracey Island* powers past Brock on the West Coast main line while working the 6J37 Carlisle Yard–Chirk Kronospan loaded log train. This service is now ran by Colas Rail and is sent via the very scenic Settle & Carlisle Railway. It was photographed here on 23 August 2007.

Owned by Virgin Trains, but holding Arriva Trains Wales livery, is No. 57315, seen here with Pendolino No. 390005 awaiting departure at Holyhead. The train was the 1A55 to London Euston, which would run to Crewe where the Class 57 would detach from the train. It was seen here on 25 September 2010.

Following on from the previous photograph, here we can see No. 57315 start its drag with No. 390005 as it screams out of Holyhead with the 1A55 to London Euston. It was seen here on 25 September 2010.

West Coast Railways No. 57601 leads the 1Z35 Carlisle–Gloucester rail tour past Taylors Bridge, Clitheroe, on 16 April 2011. No. 47826 was on the rear of the train.

Direct Rail Services Nos 57003 and 57007 head the 6K05 Carlisle North Yard–Crewe Basford Hall departmental train pass Taylors House at Clitheroe. Seen here on 10 March 2014. The train can be worked by any DRS locomotives such as Classes 20, 37, 57, 66 or 68, but it's normally a Class 66, unless there isn't one available.

On hire to GBRF, but now owned by Direct Rail Services, is Network Rail yellow No. 57306. It was seen here working the 6G31 Oubeck goods loop–Crewe Basford Hall departmental service. Seen here on 9 February 2014.

DRS Thunderbird No. 57309 *Pride of Crewe* passes Taylors House at Clitheroe while in charge of the Virgin Trains 0P99 Carlisle–Preston route refresher. It was seen here on 17 October 2013.

Direct Rail Services Nos 57007, -003 and 37607 are seen passing through Clitheroe station on Friday 14 March 2014. The working was the 0K05 Carlisle Kingmoor Sidings DRS–Crewe Basford Hall light engine move. The working is normally the 6K05 from Carlisle Yard but, due to Network Rail having no traffic, DRS used the path for a locomotive move back to Crewe.

Ex-Virgin liveried No. 57304 *Pride of Cumbria* powers past Taylors House in Clitheroe while in charge of the 0C99 Preston–Carlisle Virgin Trains route learner/refresher. It was seen here on Monday 16 June 2014.

Now owned by WCRC (West Coast Railways Company), No. 57006 was seen here passing Didcot while in charge of the very lightly loaded 4O18 05.03 Lawley Street–Southampton MCT container service. Seen here on 11 August 2005.

Nos 57004 and 57009 absolutely thrash past Brownhill after just climbing to the top of the dreaded Langho bank while in charge of the week daily 6K05 Carlisle North Yard–Crewe Basford hall departmental service. It was seen here on Monday 28 April 2014.

What is normally a departmental service became something very bizarre on Thursday 10 April 2014. Here we see Direct Rail Services No. 57004 pass Taylors House in Clitheroe while in charge of the 6K05 Carlisle Kingmoor Sidings DRS–Crewe Basford Hall.

On 9 July 2014 Direct Rail Services ran a test run for the new FNA nuclear flask wagons. It was seen here on the West Coast Mainline at Brock running in top-and-tail formation, with No. 57010 up front and classmate No. 57002 on the rear. The working was the 6Z28 Carlisle Kingmoor Sidings DRS–Derby RTC via the Cumbrian coast.

Ex-Arriva Trains Wales liveried No. 57313 is now seen working for West Coast Railways as it approaches Whalley train station, just after crossing over the forty-eight-arched Whalley Viaduct. The Class 57 was standing in for No. 60009 *Union of South Africa*, which couldn't work the train due to Network Rail having a steam ban. The working was the 1Z21 Rail Touring Companies 'Cumbrian Mountain Express' and was seen here on Saturday 16 August 2014.

No. 57308 *Tin Tin* and No. 390051 await departure from Holyhead with the 1A55 to London Euston. Seen here on 18 June 2011.

Virgin Trains No. 57316 hauls No. 390009 on the 1D83 London Euston–Holyhead drag move as it passes through Rhyl. It was seen here on 19 March 2011.

Running nearly an hour late, Nos 57012 and 57008 pass through the Staffordshire countryside at Slindon while working the 6U77 Mountsorrel–Crewe Basford Hall loaded IOA ballast boxes. Seen here on 20 June 2014.

After arriving into Preston station working the 6J37 Carlisle Yard–Chirk loaded log service where it was cancelled. The locomotive No. 57007 was on hire to Colas Rail from Direct Rail Services and is seen here waiting to depart after just running round its train. The train worked back to Carlisle as the 6Z37. Seen here on 7 January 2010.

West Coast Railways No. 57315 sits at Preston waiting to depart with the very early running 5V42 Carnforth Steamtown–Southall TC stock move. Seen here on 18 September 2014.

Due to a shortage of Class 66 locomotives, Direct Rail Services ran the daily 6K05 Carlisle North Yard–Crewe Basford Hall departmental service, with a pair of the finest Class 57s, for several weeks. Here we see Nos 57007 and -003 pass Taylors Bridge in Clitheroe with a very lightly loaded train on Tuesday 11 March 2014.

Virgin Trains' No. 57315 motors past Docker in the Cumbrian hills of the West Coast mainline while in charge of the 6J37 Carlisle Yard–Chirk loaded log train. It was seen here on 2 May 2008.

No. 57313 heads the 1Z69 Carlisle–Hereford, the Cumbrian Mountain and Coast Statesman, as it passes the outskirts of Clitheroe. No. 57316 brings up the rear on 6 April 2013.

The Class 57s are nicknamed the 'body snatchers'; here we see No. 57307 *Lady Penelope* power past Taylors House at Clitheroe with the 0C99 Preston–Carlisle route learning run on behalf of Virgin Trains. Staff practise the route in readiness for the diversions to commence. It is seen here on Wednesday 23 July 2014.

Direct Rail Services-liveried No. 57304 *Pride of Cumbria* just starts its journey as it departs Preston PCD sidings with the 0C99–Carlisle Virgin route refresher and learning run. It is routed via the Settle & Carlisle line and was seen here on Tuesday 17 June 2014.

No. 57305 *John Tracey* drags an unidentified Class 390 Pendolino set down the Ribble Valley Line as it passes Gisburn while working the 1M81 09.18 Glasgow Central–London Euston service. The Class 57 will be detached from the train at Preston, where the train will carry on as normal to London. Seen here on 12 May 2007.

Virgin Trains No. 57308 *Tin Tin* powers past the old signal box at Gisburn on the Ribble Valley line while in charge of the 1M87 13.10 Glasgow Central–London Euston Pendolino drag. It was seen here on 15 April 2007.

Newly painted No. 57010 rounds the corner at Oxcliffe Road while in charge of the 6C51 Sellafield BNF–Heysham nuclear power station flask train. It was seen here on 26 February 2015. Class 37 No. 37218 was on the rear of the train.

West Coast Railways ran the 1Z44 Leicester–Carlisle The Hadrian rail tour via the Copy Pit and Clitheroe. It was seen here with No. 57601 leading the tour and No. 47245 on the rear, which was out of sight. It was photographed here on 10 October 2015.

On 29 March 2007, Freightliner Reliance No. 57006 was sitting in the yard in Ipswich awaiting its next duty on an intermodal service from the Port of Felixstowe. The locomotive is now owned by West Coast Railways at Carnforth.

Sitting in the Down and Up goods loop at Carnforth is Direct Rail Services Nos 57004 and 002 while being in charge of the 6K73 Sellafield BNF–Crewe Coal Sidings DRS flask service. It was seen here on 10 August 2015.

No. 57303 *Alan Tracey* powers past Red Bank near Newton-Le-Willows while in charge of the 6J37 Carlisle Yard–Chirk loaded log train. It was seen here on 2 April 2009.

No. 57309 is photographed hauling No. 390047 on the 1D83 London Euston–Holyhead, minus Virgin branding and its *Brains* nameplates. The train is departing Rhyl on Saturday 24 March 2012.

DRS body snatchers Nos 57007 and -012 pass through Clitheroe with the weekday running 6K05 Carlisle North Yard–Crewe Basford Hall departmental service. It was seen here on Wednesday 19 March 2014.

No. 57314 passes the very bleak Blea Moor signal box while in charge of the 13.10 1M87 Glasgow Central–London Euston Pendolino drag service. Seen here on 4 March 2006.

Direct Rail Services No. 57003 drags Nos 66302, 37607 and 37601 past Acton Cliff while working the 6K74 09.09 Sellafield BNF–Crewe Coal Sidings DRS flask train. It was seen here running ninety minutes early on the 30 January 2016.

West Coast Railways Class 57 No. 57314 arrives into Hellifield goods loop after working the 0Z57 from Carnforth Steamtown. Seen here on 28 July 2015.

No. 57310 *Kryano* powers past Chatburn on the Ribble Valley Line while in charge of the 1M87 13.10 Glasgow Central–London Euston Pendolino drag. It was seen here on 11 May 2008.

Nos 57004 and 57011 have just crossed over the twenty-four-arched Ribblehead Viaduct while being in charge of the 6K05 Carlisle North Yard–Crewe Basford Hall departmental service. Seen here on 29 April 2014.

No. 57304 *Gordon Tracey* powers past Settle Junction while working the diverted 1M19 15.05 Glasgow Central–London Euston service. It was seen here on 28 May 2008.

Ex-Arriva Trains Wales blue No. 57316 was acting as a Thunderbird, but running this time for its new owners West Coast Railways. The locomotive was sent from Carnforth Steamtown as the 1Z99 to Newark Northgate to rescue LNER A4 Class 4-6-2 No. 60009 *Union of South Africa* and LNER K4 Class 2-6-0 No. 61994 *The Great Marquees* due to a high fire risk being placed by Network Rail. The train was seen here arriving into Hellifield goods loop running some seventy minutes late. The working was the 05.10 5Z90 Dereham–Carnforth Steamtown and was seen here on Wednesday 2 July 2014.

No. 57308 *Tin Tin* rounds the curve into Chatburn on the Ribble Valley Line while working the 1M87 13.10 Glasgow Central–London Euston Pendolino drag. It was seen here on 2 June 2007. The locomotive still acts as a Thunderbird to this day but is now owned by DRS and is named *County of Staffordshire*.

Freightliner-owned No. 47830 *Beechings Legacy* is seen on hire to Rail Operations Group (ROG) as it thrashes past Scorton while in charge of the 06.00 Kilmarnock (Barclay Sdgs)–Laira T&RSMD FGW stock move. Seen here on Saturday 9 April 2016.

With its radiator shutters wide open, No. 47805 storms towards Chester with 1Z47 11.44 Coventry–Coventry, via Bescot, Stafford, Crewe, Chester, Wrexham, Shrewsbury and Wolverhampton, Northern Belle Sunday lunch trip. Forty-seven-years old and still going strong, No. 47805 was released into traffic from Brush Falcon works on March 18 1966. Its classmate No. 47828, also forty-seven years old but a Crewe Works-built example, brings up the rear of the train.

Freightliner No. 47830 *Beechings Legacy* sits at Leeds Midland Road depot, awaiting its next duty of call. It is seen here parked right up to the buffer stops, while Class 66 No. 66596 sitting behind the locomotive. It is pictured here on Friday 22 April 2016.

Another 57/3 Thunderbird stripped of its names and Virgin branding in readiness to be handed over, No. 57304 ex-*Gordon Tracy* screams through the Cheshire countryside at Calveley. It is working 1D83 08.50 London Euston–Holyhead, with No. 390035 in tow.

Virgin trains No. 57305 sits at Warrington Bank Quay train station while in charge of the 09.50 Manchester Piccadilly–Holyhead locomotive-hauled passenger train. It was photographed here on 15 February 2006.

No. 57306 *Jeff Tracy* climbs through the lovely countryside of the North Yorkshire Dales after just crossing over the border out of Lancashire at Newsholme while in charge of the diverted 1S45 08.45 London Euston–Glasgow Central service. It was seen here on 4 March 2006.

No. 57307 *Lady Penelope* thrashes over Whalley Viaduct while in charge of the diverted 1M81 09.18 Glasgow Central–London Euston passenger service. It was seen here in the snow on 4 March 2006.

D1501 (No. 47402) sits at Rawtenstall on the East Lancashire Railway after just arriving with the first service from Bury. The locomotive is seen in two-tone green livery and is based at the East Lancashire Railway. It was seen here on 9 January 2010.

Direct Rail Services No. 47832 *Solway Princess* is seen working the 0Z47 Carlisle Kingmoor sidings–Carlisle Kingmoor sidings DRS route-learning move. It was seen here passing through Clitheroe on 6 August 2008. The loco is now owned by West Coast Railways, who are based at Carnforth.

InterCity-liveried No. 47826 screams past Rishton reservoir while working the 0Z47 Healey Mills–Carnforth Steamtown Class 47 move. In this consist were Nos 47760, 47776, 47786, 47792. The locomotives had been bought by West Coast Railways from EWS. They were noted here on 31 March 2007.

Ex-Arriva Trains Wales blue No. 57316 was acting as a Thunderbird, but running this time for its new owners West Coast Railways. The locomotive was sent from Carnforth Steamtown as the 1Z99 to Newark Northgate to rescue LNER A4 Class 4-6-2 No. 60009 *Union of South Africa* and LNER K4 Class 2-6-0 No. 61994 *The Great Marquess* due to a high fire risk being placed by Network Rail. The train was seen here arriving into Hellifield goods loop running some seventy minutes late. The working was the 05.10 5Z90 Dereham–Carnforth Steamtown and was seen here on Wednesday 2 July 2014.

ALSO AVAILABLE FROM AMBERLEY PUBLISHING

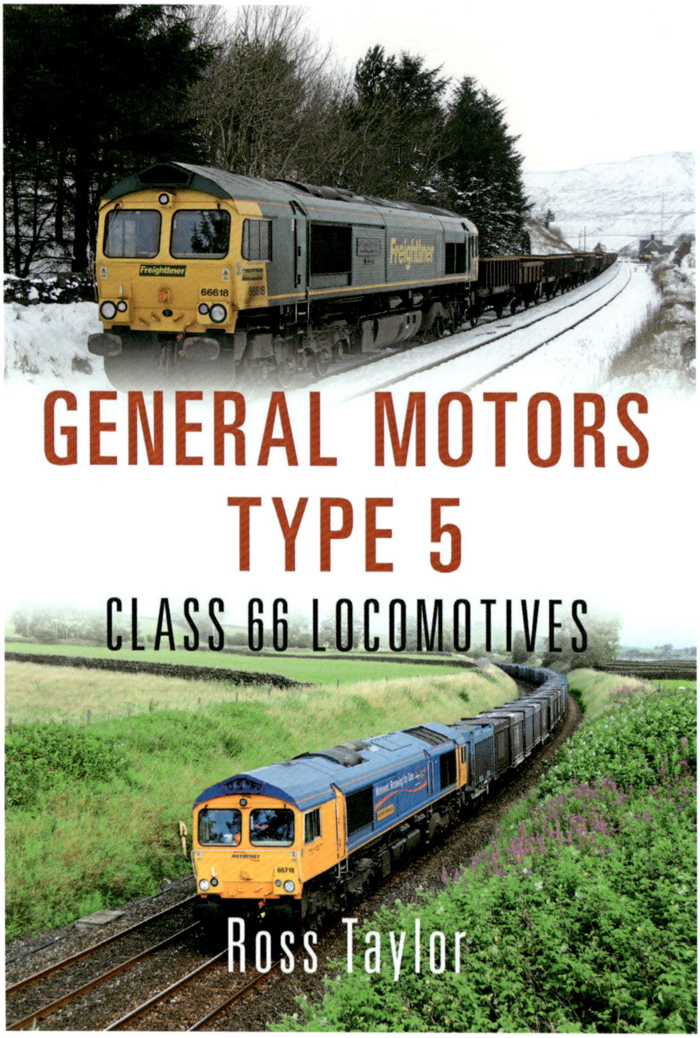

General Motors Type 5 Class 66 Locomotives

Ross Taylor

Ross Taylor explores the fascinating world of Class 66 locomotives.

978 1 4456 4850 7
96 pages, illustrated throughout

Available from all good bookshops or order direct
from our website www.amberley-books.com